YOUR KNOWLEDGE HAS VALUE

- We will publish your bachelor's and master's thesis, essays and papers

- Your own eBook and book - sold worldwide in all relevant shops

- Earn money with each sale

Upload your text at www.GRIN.com and publish for free

Bibliographic information published by the German National Library:

The German National Library lists this publication in the National Bibliography; detailed bibliographic data are available on the Internet at http://dnb.dnb.de .

This book is copyright material and must not be copied, reproduced, transferred, distributed, leased, licensed or publicly performed or used in any way except as specifically permitted in writing by the publishers, as allowed under the terms and conditions under which it was purchased or as strictly permitted by applicable copyright law. Any unauthorized distribution or use of this text may be a direct infringement of the author s and publisher s rights and those responsible may be liable in law accordingly.

Imprint:

Copyright © 2015 GRIN Verlag, Open Publishing GmbH
Print and binding: Books on Demand GmbH, Norderstedt Germany
ISBN: 9783668518520

This book at GRIN:

http://www.grin.com/en/e-book/373291/aviation-safety-regulatory-framework-technology-contingency-plan

Alfhonce Michael

Aviation safety. Regulatory framework, technology, contingency plan

GRIN Publishing

GRIN - Your knowledge has value

Since its foundation in 1998, GRIN has specialized in publishing academic texts by students, college teachers and other academics as e-book and printed book. The website www.grin.com is an ideal platform for presenting term papers, final papers, scientific essays, dissertations and specialist books.

Visit us on the internet:

http://www.grin.com/

http://www.facebook.com/grincom

http://www.twitter.com/grin_com

Content

AVIATION SAFETY 2
REGULATORY FRAMEWORK 2
TECHNOLOGY 7
CONTINGENCY PLAN 8
References 9

AVIATION SAFETY

REGULATORY FRAMEWORK

Air transport has grown tremendously over the last centuary. The launch of the Jumbo jet was the game changer in commercial air transport. The industry has expanded greatly since then and today it is a multi-billion industry employing thousands of people and providing transport services to millions yearly. Despite the growth, Air transport has been faced with increasing security and safety concerns. This is evident in the large number of air crashes recorded every year as well as the bombing of commercial airliners.Inorder to ensure safe air travel, numerous organizations and institutions have come on board to develop rules, regulations and standards on safety. Regulatory framework and safety requirements have been built up over the years and are continuously been enhanced to address emerging issues as pertains air security and safety. These organizations include International Civil Aviation Organization (ICAO), European Aviation Safety Authority (EASA), National Transport and Safety Board (NTSB).Civil Aviation Authority (CAA) and Federal Aviation Administration (FAA).

Safety is key in Avaition.The industry is built on safety. There are there layers in safety regulation in Aviation. They are International, Regional and National regulatory arrangements. International regulatory requirements are addressed by ICAO.The ICAO is an agency of the United Nations and was established in 1944 through a convention on International Civil Aviation (Lavenex 2008, 938). The organization develops standards that cover all aspects of aviation including safety. Through its Standards and Recommended Practices (SARPS), it provides the foundation of all safety regulations at a global scale. It oversees the development of safety regulatory framework by Member states through Universal Safety Oversight Audit Programme (USOAP). In recent years, ICAO requirements have been extended to require the implementation of a formal safety management by aviation service provider organizations as well as aircraft operators (Mclay 2008, 107). Regional regulatory arrangements is a layer that cedes National Regulatory functions to supra-national agencies. A good example is EASA. It was established in 2003 as an agency of the European Union. Its current mandates include initial certification determination of airworthiness of aircraft and related products as well as the approval of organizations involved in the design, manufacture and maintenance of aeronautical products. Further, it certifies personnel and organizations involved in operating an aircraft (Pietre 2010, 55). Regulations of the EASA are reached upon by member states of the EU. Finally, National regulatory requirements are promulgated in National legislations by the designated state authorities. These regulations are objective based on individual country.

Security is another critical aspect in Aviation. Aviation security is broad and includes all ground and air operations aimed at making Aviation secure. In the U.S.A, Aviation security has been governed by the FAA with significant private sector participation. Aviation security in the country came under intense scrutiny in the aftermath of September 11th attacks when terrorist hijacked four airliners crashing two into the World Trade Centre in New York and one near the Pentagon in Washington. Since the attacks, the FAA has adopted a multi-layered security regulation (Stewart 2008, 143). Canada also transferred its Aviation security to a federal agency, The Canadian Air Transport Security Authority (CATSA).The Multi-layer approach is based on the principle that if one layer is breached, the second will hold firm. Another body charged with ensuring security in Aviation is the NTSB.This is the body that investigates air crash accidents involving all commercial aircraft and military aircrafts in the U.S.It was established in 1967 as an independent investigative agency by congress/It is given priority in investigations into causes of aircrashes.It is the only body mandated to determine causes of accidents (Sweet 2008, 100).

However, it is not given exclusive power. It is supposed to ensure appropriate participation by other agencies. It is mandated to ensure appropriate information developed about accidents is exchanged in a timely manner. In the U.K, Aviation security regulation is done by the CAA.It develops detailed regulatory requirements and guidance to support Aviation security policy (Pietre 2000, 58).

Three theories have been developed to explain the causes of aircrashes.They are referred to as the Accident causation models. The three models are simple linear, complex linear and complex non-linear models.sequntial events. The simple linear model presumes accidents are the result of a sequence that involves social, environmental and individual factors and mechanical or physical hazards (Pietre 2010, 60). The complex linear model assumes accidents are the result of a number of unsafe conditions where the flight crew is at risk. Finally, the complex non-linear model states that accidents are caused by mutually interacting variables in real time environments.It states accidents can be avoided by understanding the interaction of these factors (Lavenex 2008, 940). The three theories above point the fact that human beings are at the center of air crashes. Studies have revealed that human error is responsible for more than 70% of air crashes globally. These statistics show the integral role that human play in the aircraft. The Human Factor involves a combination of factors including psychological state, environmental factors, emotional state as well as mental factors affects a flight crews decision making as they interact with an aircraft's systems (Mclay 2008, 125).

One of the Aviation incidents where the human factor played a key role was in the near crash of China airlines flight 006 in 1985.The flight was a routine one from Taipei to Los Angeles. The plane was a Boeing 747.Captain Ho, an experienced pilot was in charge of the plane together with three other crew. As the plane neared the coast of California, it experienced light turbulence.

Soon after, the 4th engine began giving a weak thrust. The flight engineer immediately throttled the engine up but it did not respond (Lavenex 2008, 947). The Captain immediately called the Oakland control center and requested that he be assigned a lower altitude to retry the procedure. As the Captain tried to restart the plane, it began tilting to the left and begins falling from the sky (Lavenex 2008, 944). With thirty minutes left before the plane plunges in to the Pacific, the captain sees the horizon and begins to level the plane pulling it out of the dangerous nose dive. On landing, damage to the plane is evident. The door to the landing gear is torn off and there are damages to the wings. It was only a miracle that the plane landed safely (Stewart 2008, 148). The human factor has been identified as critical in addressing aviation safety. A lot of research has been launched to address the same. The NTSB takes the flight data recorder to Washington for analysis into the cause of the near accident. The first instrument to be investigated is the ADI which levels the plane. It is found to be working. Three engines are also found to be in order contrary to what the flight engineer had said (Pietre 2010, 58).

The investigation found that the three engines responded slowly when throttled up because of the cold temperatures at 9000 feet. After concluding that there was no problem in the planes mechanical components, attention shifts to the pilot and his crew. The four crew are interviewed. The investigators realize contradiction into what the flight engineer was saying and what they had discovered. He says that the three engines were off but investigations showed they were functioning all through (Mclay 2008, 124). Investigators later realize that the engineer failed to notice that the Captain left the engine thrust to idle to decrease fall of the plane from the sky. On interviewing the Captain, the investigators discover that he had a hard time falling to sleep during the break he and his copilot had. This is likely to have affected his decision making. He is also observed to have over relied on the autopilot (Pietre 2010, 54). This is the reason he failed to step on the rotor when the plane began turning. Instead, he was concentrating on the altitude indicator.Invstigators conclude the near accident was as a result of a series of wrong decisions made by the crew. They recommend for the redesign of an airplanes auto-pilot systems to let the Captain have control as much as possible (Lavenex 2008, 940).

From the above incident, it is evident that human factor plays a crucial role improvement of aviation safety. From the incident, investigators recommended that the auto-pilot should be redesigned to give the pilot that everything more control. This is after realization it could fool the pilot into believing everything in the aircraft was fine when it was not (Pietre 2010, 61). Consequently, Boeing restructured their automated pilots and set the appropriate degree of automation. Its flight decks now have intuitive, easy to use systems. The systems have visual and tactile motion cues to minimize potential confusion. Just like in the China Airlines, human factor air crashes and near accidents have led to recommendations that have improved flight deck design and communication making aircrafts safer (Sweet 2008, 157). To further bolster safety and security in Aviation, international organizations, and air navigation service providers put the Safety Management Systems (SMS) agenda on the table in the 1990's.SMS's are objective result oriented standards designed to allow air Aviation service providers to integrate safety management practices in their individual operation models.

TECHNOLOGY

One of the security requirement by the FAA is the use of security scanners in all airports in the U.S.A.ON December 5, 1972, it issued an emergency rule for immediate screening of passengers and their baggage. Screening programs enable carriers to implement a security program culpable of preventing introduction of weapons and explosives onboard aircrafts. Today, all Airports in the U.S screened before boarding aircrafts. They employ the use of hand-wand metal detector. Meanwhile, passenger's baggage is run on a conveyor belt for inspection by X-Ray equipment (Pietre 2010, 54). Full body scanners and iris scanners are other new technology in the field. I believe that the security measures and technology being used is not sufficient. Most of the detectors in use can only detect metal objects and cannot detect plastic explosives. There is research to come up with such detectors (Lavenex 2008, 947). This leaves a gap which terrorists can use. In addition to the technology aspect, other important aspects that will determine the success or failure of the technology include proper decision making and judgment by security personnel who screen passengers and acceptability of the technology amongst members of public (Mclay 2008, 123). If security personnel are unable to use the technology at their disposal well, bombs will eventually find way into aircrafts. This calls for proper training. Interms of acceptability of the new technology, issues such effects to health and privacy are of concern and should be addressed.

CONTINGENCY PLAN

Air crashes can occur at any time. Most air crashes occur during takeoff and landing. It is imperative that airports be prepared all times for these crashes. This involves having a fully equipped efficient emergency unit. One of the air crashes where a significant number of lives were saved due to efficient response was Delta Flight 191 which crash landed at Dallas-Fort Worth Airport (Pietre 2010, 64). The twenty nine survivors were rescued by emergency teams from the Airport before the plane burst into falmes.Based on this, it is evident efficient emergency services can save lives in air crashes. My contingency plan to tackle an accident at an airport involves the following measures.

1. Quickly determine the magnitude of the accident to know the resources to deploy to effectively deal with the accident.

2. Quickly mobilize and dispatch a rescue team comprising of medical personnel, firefighters and a rescue team.

3 Notify nearby hospitals of the incident so they can create room for survivors.

4. Immediately set up a makeshift hospital at the site in case there are people without serious injuries.

5. Quickly establish if there are people thrown out of the cabin and attend to them.

6. Look into ways of getting into the plane if it is not broken up and isn't on flames.

7. If the plane is on flames, let the fire team immediately move in with their equipment to put out the fire as they try to rescue survivors.

References

Lavenex, S., 2008. A governance perspective on the European neighbourhood policy: integration beyond conditionality?. *Journal of European public policy*, *15*(6), pp.938-955.

McLay, L.A., Jacobson, S.H. and Kobza, J.E., 2008. The tradeoff between technology and prescreening intelligence in checked baggage screening for aviation security. *Journal of Transportation Security*, *1*(2), pp.107-126.

Piètre-Cambacédès, L. and Chaudet, C., 2010. The SEMA referential framework: Avoiding ambiguities in the terms "security" and "safety". *International Journal of Critical Infrastructure Protection*, *3*(2), pp.55-66.

Stewart, M.G. and Mueller, J., 2008. A risk and cost-benefit assessment of United States aviation security measures. *Journal of Transportation Security*, *1*(3), pp.143-159.

Sweet, K., 2008. *Aviation and airport security: terrorism and safety concerns*. CRC Press.

YOUR KNOWLEDGE HAS VALUE

- We will publish your bachelor's and master's thesis, essays and papers

- Your own eBook and book - sold worldwide in all relevant shops

- Earn money with each sale

Upload your text at www.GRIN.com
and publish for free